Rainwater Harvesting for Irrigation

Discover Everything You Need to Master Rainwater Harvesting in Your Garden or Farm | Fast, Easy and Safe Solutions to Create a Clean Water Source and Save Money

Melanie J. Davis

© Copyright 2023 by Melanie J. Davis All rights reserved.

All rights reserved. No part of this book may be reproduced in any form without permission in writing from the author. Reviewers may quote brief passages in reviews.

While all attempts have been made to verify the information provided in this publication, neither the author nor the publisher assumes any responsibility for errors, omissions, or contrary interpretation of the subject matter herein.

The views expressed in this publication are those of the author alone and should not be taken as expert instruction or commands. The reader is responsible for his or her own actions, as well as his or her own interpretation of the material found within this publication.

Adherence to all applicable laws and regulations, including international, federal, state and local governing professional licensing, business practices, advertising, and all other aspects of doing business in the US, Canada or any other jurisdiction is the sole responsibility of the reader and consumer.

Neither the author nor the publisher assumes any responsibility or liability whatsoever on behalf of the consumer or reader of this material. Any perceived slight of any individual or organization is purely unintentional.

Table of Contents

INTRODUCTION 5

The Science of Rainwater Harvesting 7

How Rainwater Is Collected and Stored 7
How rainwater is filtered and treated 10
The quality of harvested rainwater 13

The Benefits of Rainwater Harvesting for Gardeners 15

Why Use Rainwater to Water My Gardner 15
The Choice Behind Rainwater Harvesting 16
All the Numerous Advantages 17

Getting Started with Rainwater Harvesting 20

Determining Your Water Needs 20
Choosing the right system for your garden 22
Equipment & Supplies Needed 26

Setting Up Your Rainwater Harvesting System 29

How to install a rainwater collection system 29
How to Properly Position & Maintain The System 33
Troubleshooting 36

DIY Rainwater Harvesting Projects 39

Building Your Own Barrel 39
Building Your Own Cistern 41
Building Your Own Underground Storage Tank 43
Building Your Own Irrigation System 45
How to Setup The Perfect Rain Garden 47

Using Harvested Rainwater in Your Garden 50

Best practices for using harvested rainwater 50
How to Incorporate Rainwater Harvesting Into Your Gardening Practices 53
Tips for maximizing the benefits of harvested rainwater for your garden 55

Choosing the Right Plants for Your Rainwater Garden 58

Well suited 58
Designing a Rain Garden To Maximize Water Usage And Plant Health 61
Tips for Maintaining Your Rain Garden 64

Rainwater Harvesting for Urban Gardening 67

All the Benefits 67
Overcoming Challenges 70
Innovative Ways To Use Rainwater In Urban Gardens 71
How to Setup Your Rainwater Harvesting System 74

Rainwater Harvesting in Different Climates 78

The importance of climate 78
Rainwater Harvesting in Arid Climates 80
Rainwater Harvesting in Rainy Climates 81
Adapting rainwater harvesting techniques to different climate conditions 83

Rainwater Harvesting for Farming and Agriculture 87

Benefits of Rainwater Harvesting for Agriculture 87
Different Types of Rainwater Harvesting Systems for Farming 90
How to Maximize the Use of Harvested Rainwater for Crops and Livestock 92
How to Setup Your Rainwater Harvesting for Farming & Agriculture 95

Legal and Regulatory Considerations 98

Legal considerations for rainwater harvesting 99
Regulations in Different States and Regions 102
How to Ensure Compliance With Local Regulations 105

CONCLUSION 107

INTRODUCTION

Gardeners, welcome to the world of rainwater harvesting! If you enjoy gardening and are concerned about the environment, you're in for a treat. Water is the essence of life, and it is essential to the health of a garden. However, with growing concern about water scarcity and the need to conserve our natural resources, learning how to collect and use rainwater effectively has never been more important.

Rainwater harvesting is a long-standing and environmentally friendly solution. Rainwater harvesting techniques have been used by people from ancient civilizations in the Middle East to modern-day Australia to collect and store water for a variety of purposes. This self-sustainable practice has grown in popularity among gardeners in recent years as a practical and environmentally friendly way to water their plants. It not only reduces the demand on municipal water supplies, but it also provides a free and natural water source for your garden. *But where do you begin with rainwater harvesting?* This is where this detailed guide comes in!

We'll take you on a tour of the world of rainwater harvesting for gardeners in this book. We'll begin by looking at the science of rainwater, such as how it forms and why it's beneficial to your garden. Following that, we'll look at the various types of rainwater harvesting systems available, ranging from simple barrel systems to more complex setups capable of handling large volumes of water. We'll go over the benefits and drawbacks of each type of system so you can choose the best one for your needs.

Once you've decided on a system, we'll walk you through the installation process, which includes determining the best location for the system, setting it up, and connecting it to your garden. We'll also go over some maintenance and troubleshooting techniques to keep your system running smoothly year after year.

But wait, there's more! We'll also give you tips on how to use rainwater for all of your gardening needs, such as irrigation, watering, and fertilization. You'll learn how to design a beautiful and sustainable water-wise garden.

This book is your ultimate resource for understanding the benefits of rainwater harvesting, exploring the various types of harvesting systems available, and learning how to use rainwater for all of your gardening needs, whether you're an experienced gardener or just starting out. So, let's get started and see how rainwater harvesting can transform your garden and help you contribute to a more sustainable world!

Have a good read,

Melanie J. Davis

CHAPTER 1

The Science of Rainwater Harvesting

It's crucial to examine the physics behind rainwater harvesting in order to fully comprehend its advantages and workings. In this chapter, we'll delve into the intriguing world of rainwater and its characteristics, such as how it forms, why it's good for your garden, and practical methods for gathering and storing it. You'll have a thorough grasp of the science behind rainwater harvesting by the end of this chapter, which will allow you to choose your harvesting system wisely and get the most out of it for your garden. So let's delve in and discover the science behind rainwater harvesting!

How Rainwater Is Collected and Stored

Although it is a scarce and priceless resource, water is crucial for plant growth. It is worthwhile to preserve and collect it in order to get the most use out of it.

As long as they have gutters and a drainage pipe, houses, greenhouses, and other garden structures can have rainwater collected from their roofs. In DIY stores, sturdy plastic containers can be purchased. Water will be accessed through a spigot that is built into the container's base, so a sturdy stand should be made, even if it has to be made out of a few bricks. Earthenware or beehive-shaped containers, which are more expensive, are a desirable option, as are recycled wooden barrels. It may be cost-effective to include rainwater storage in the construction of new homes as climate change models predict that an increasing proportion of rain will fall during the winter months (such as a large tank in the garden). A system for recovering rainwater (or stormwater) consists of:

- A storage tank;

- One or more downpipe filters;

- A water withdrawal pump.

The water brought here by downspouts on the roof is collected, filtered, and stored in the rainwater tank before being circulated again through a network of pipelines and withdrawal pumps. All tanks used to collect rainwater are very identical to one another. Size and material are the only distinctive characteristics.

The size of the tank relies entirely on its capacity; there are small, medium, and large tanks, each of which can hold dozens to hundreds of gallons of water. The decision between the two is influenced by several variables:

- The water needs of the house;

- The type of system (for domestic use or irrigation);

- The rainfall of the area;
- The collection surface;
- The material from which the area is composed.

The cistern must have a drainage system or be linked to the public sewer, and it must be put in the yard or another area close to the house or building it is intended to serve. This is done to promote rainfall runoff.

The substance used to construct the tank is another consideration. Rainwater containers can be positioned outdoors, either underground or above ground. They must therefore be constructed of materials that are ideal for coming into touch with water without oxidizing because they are continuously exposed to the elements, atmospheric agents, and potential temperature changes. This is why PVC or polyethylene, materials that do not deteriorate or change over time, are used to make the majority of rainwater collecting containers.

There are some crucial steps to take in order to properly collect and store rainwater for your plants. In order to prevent the water from becoming stagnant or even gathering debris that contributes to water pollution, it is crucial that the container that will house the water collection is easily resealable. Also keep in mind that standing water that is not properly covered makes an excellent home for bothersome insects like mosquitoes, which thrive especially in the summer. Take advantage of rainy days to gather as much water as you can if you have a lot of plants or if it doesn't rain often. To accomplish this, try piping water into the container of your choice while it is flowing from the gutter. Use the collected water within a day or two to avoid it starting to stagnate if the container you chose for water collection does not close securely or is left open.

A rainwater harvesting system is an excellent method to reduce your reliance on the city's water supply. This is so that the water that falls from the heavens can be captured, stored, and used for irrigation or other purposes inside the house thanks to the collection tanks.

How rainwater is filtered and treated

Is it possible to successfully purify rainwater? You may have asked yourself this question a few times as you watched this priceless resource fall from the sky and considered your needs for conserving water. The topic of rainwater purification and the potential for using the blue gold that rains from the sky to supply our water needs will be the main points of this chapter.

It is possible to drink rainwater, but only in extremely dire circumstances. In general, it is always advised to purify rainwater before using it for purposes that call for potable water, such as personal hydration:

- **Pollution:** Water released from clouds travels through an atmosphere that, depending on the location, may also be heavily contaminated and loaded with all the toxic substances suspended in the air before it reaches the ground.

- **Water Quality:** Mineral salts and limestone are absent from rainwater. It is not suitable for individual consumption because it lacks the nutrients required by the human body, making it more suitable for domestic uses (maintenance of household appliances, washing of dishes and floors, watering of gardens and vegetable gardens) or industrial uses.

- **Acidity:** Rainwater has a PH of about 5.6, making it slightly acidic, but industrial emissions can raise this value to be between 4 and 4.6. Drinking such acidic water is certainly not healthy and it would be best to avoid its consumption.

Therefore, should we stop using this resource that rains from the sky, or is it possible to use a process to make rainwater potable? As previously mentioned, rainwater is never microbiologically pure, so in order to use it for potable purposes, it must be cleaned of all pathogenic elements, bacteria, viruses, heavy metals, and other contaminants.

Boiling rainwater is the simplest way to purify it. The germs and bacteria can be eliminated by heating the water to 100° C and letting it boil for 10 minutes.

After being cleaned, the water must be put through special filters, but only after being allowed to decant for at least 12 hours away from impurities. This is a system that can be used occasionally to treat small amounts of water, but it is by no means a comprehensive approach. On the other hand, you can use more organized filtered systems and facilities to purify rainwater in order to make large volumes potable:

- **Ceramic filtration**: This material's microporosity has the ability to rid the water of pathogenic substances like salmonella, sediments, and other impurities that it picks up during its fall to the ground.

- **Uv Ray Depuration**: UV-based disinfection is particularly effective at getting rid of bacteria and viruses without the use of chemicals.

- **Chlorine Depuration:** We put this, one of the most popular rainwater filtration and purification systems, to the test every time we jump into a pool, confident that the water can be consumed without worry.

Should you not be interested in setting up systems to purify large volumes of rainwater, there are safe and quick alternatives. Here are some quick and safe ways to purify small amounts of rainwater:

- **Filtering:** Larger debris and particles can be removed from rainwater by filtering it through a fine mesh or cloth. To eliminate bacteria, viruses, and other impurities from water, you can also use a water filter made for camping or emergency situations.

- **Chemical treatment:** Tablets containing chlorine or iodine are frequently used to purify water in emergencies. Follow the dosage and waiting period recommendations provided by the manufacturer.

- **Solar disinfection:** This technique, also referred to as SODIS, uses sunlight to purify water. Water should be added to a clear plastic bottle, which should then be left in the sun for at least six hours. Bacteria and viruses are killed by the sun's UV rays.

It's important to keep in mind that these techniques might not completely remove all impurities and are only effective for small amounts of water. It is best to have your rainwater tested by a laboratory before consumption if you are unsure of its safety.

The quality of harvested rainwater

As we've seen, rainwater is a useful resource that offers a variety of advantages, but it's crucial to think about the water's quality before using it. This chapter will examine the elements that can impact the quality of rainwater that has been collected as well as ways to guarantee that the water is suitable for its intended use.

A number of variables, such as the collection surface, storage setup, and environmental conditions, can have an impact on the quality of harvested rainwater. The type of roofing material, for instance, can have an impact on the water's quality because some materials can leach chemicals into the water. In a similar vein, the storage system needs to be well-planned and kept up to date to avoid contamination from bugs, animals, or other sources.

It's crucial to test the water frequently to make sure that rainwater collected for use is safe. Hazardous contaminants like bacteria, viruses, and heavy metals can be found through testing. You can either purchase testing kits or have a laboratory test the water. Testing the water is advised at least once a year and following any significant alterations to the collection or storage system.

The appropriate use of rainwater will depend on its quality. A lesser level of purification might be adequate for non-potable uses like watering plants or flushing toilets. However, the water must adhere to strict quality standards in order to be used for potable purposes like drinking or cooking. In order to lessen the demand on municipal water supplies, it is generally advised to use collected rainwater for non-potable purposes.

Gardeners, farmers, and homeowners can all benefit from the use of harvested rainwater. To make sure the water is secure for its intended use, the quality must be carefully taken into account. Harvested rainwater can be made a safe and dependable water source by routine testing and using the right treatment techniques.

CHAPTER 2

The Benefits of Rainwater Harvesting for Gardeners

Harvesting rainwater to water your plants is a simple and inexpensive solution. But what makes rainwater the ideal choice for watering plants? In this chapter you will find out why watering your plants with rainwater is the winning choice.

Why Use Rainwater to Water My Gardner

Owning numerous plants makes it natural for us to water them with tap water because it is the quickest and most practical solution. However, not everyone is aware that watering plants with properly collected rainwater can have numerous advantages for the plants' healthy growth. Here's why:

- Rainwater is perfect for watering delicate plants and vegetables because it is naturally soft and has fewer dissolved minerals than groundwater.

- Rainwater's mild acidity can help reduce the pH of overly alkaline soils and increase the availability of nutrients for plants.

- Finally, rainwater is much better for watering plants because it is free of chemicals that might be in municipal waterworks.

In addition to reducing your household water usage, collecting rainwater to water your plants has a positive environmental effect. Utilizing rainwater reduces the amount of drinking water that is wasted, a resource that is sadly destined to run out.

Reusing rainwater is not only the best option for having lush, healthy plants, but it also becomes the morally right thing to do in order to conserve drinking water and contribute to the preservation of our frail planet, which is increasingly experiencing harsh and dangerous droughts.

The Choice Behind Rainwater Harvesting

Rainwater collection and use is not only a practical decision, but also an ethical one. It's crucial to take into account the effects our daily actions have on the environment and other people in a world where access to clean water is becoming more and more difficult to come by. Because it lessens our reliance on municipal water supplies, which are frequently chemically treated and can be expensive, collecting and using rainwater is a moral decision. We can lower our water bills and preserve water resources for future generations by collecting and using rainwater.

Rainwater collection can also aid in reducing the effects of climate change. Finding efficient methods for managing and storing water is crucial as extreme weather events like droughts and floods become more frequent. We can lessen the amount of stormwater runoff, which can cause soil erosion, water pollution, and even flooding, by collecting and using rainwater. It also helps to foster resiliency and self-sufficiency. We reduce our vulnerability to water shortages and disruptions in the municipal water supply by collecting and storing rainwater. Having a dependable water source is crucial in emergency situations.

The decision to collect and use rainwater is ultimately morally right because it acknowledges the value of clean water as a common resource. We can help ensure that future generations have access to abundant, clean water by being accountable for our own water needs.

All the Numerous Advantages

Gardeners can benefit greatly from this practice, both financially and environmentally. In this section, we'll explore the top 4 benefits of rainwater harvesting for gardeners.

- **Reduced Water Usage for Gardening:** The reduced water use for gardening is one of the most important advantages of rainwater harvesting for gardeners. Gardeners can significantly lessen their reliance on municipal water supplies by collecting and storing rainwater. This not only lowers their water bills but also encourages resource conservation, which is important during dry spells or when there are water restrictions.

- **Cost Savings on Water Bills:** Water bill savings are another benefit of rainwater harvesting. Gardeners can reduce their reliance on municipal water sources and lower their water costs. Depending on the size of the system and the amount of water used, some rainwater harvesting systems can pay for themselves within a few years.

- **Improved Plant Health and Yield:** Rainwater is naturally softer than tap water and devoid of many contaminants like fluoride and chlorine. This makes it a great source of water for plants because it encourages healthy plant growth and maintains healthy soil. Studies have actually shown that plants watered with rainwater are typically healthier and more fruitful than those watered with tap water.

- **Reduced Environmental Impact:** Gardeners can significantly reduce their environmental impact by collecting and using rainwater. They can contribute to water resource conservation and lower the energy needed to transport and treat water by reducing their reliance on municipal water supplies. Rainwater collection can also lessen stormwater runoff, which helps stop soil erosion and water pollution.

For gardeners looking to use less water, pay less for water, and encourage healthy plant growth, rainwater harvesting is a great practice. Gardeners can significantly lessen their environmental impact and help to create a more sustainable future by collecting and using rainwater. *So why not setup your rainwater system right away and benefit from all that it has to offer?* The environment, your wallet, and your garden will all thank you.

CHAPTER 3

Getting Started with Rainwater Harvesting

Rainwater harvesting is a worthwhile endeavor if you're interested in conserving water, lowering your water bills, and encouraging a more sustainable future. But it can be overwhelming to know where to start if you're new to the practice. We'll go over the fundamentals of getting started with rainwater harvesting in this chapter, including the various types of systems that are available, how to determine your water needs, and equipment and supplies needed. The following lines will provide you with the skills and information necessary to begin harvesting rainwater right away!

Determining Your Water Needs

A critical first step in developing a sustainable garden with the aid of rainwater harvesting is determining your garden's water requirements. You must consider a number of variables that affect your plants' water needs in order to ensure that they receive the appropriate amount of water.

Let's take the example of a 20 feet by 20 feet garden where a variety of vegetables, flowers, and herbs are being grown. According to the University of California Cooperative

Extension, flowers and herbs may need less water than vegetables, which typically need 1-2 inches per week. This means that in order to provide your plants with the water they require, you will need to collect at least 500 gallons of water each month for a garden this size.

The local climate and weather patterns also play a role in how much water you need to gather. You might not need to gather as much water as you would in an area with long, dry summers if there is frequent rainfall there. You might be able to rely almost entirely on rainwater for your garden's needs if you live in a region like the Pacific Northwest, for instance, where rainfall is abundant all year long.

On the other hand, you'll need to be more methodical in your rainwater harvesting efforts if you reside in a region like California where water shortages are common. During dry spells, you might need to install a more complex system with storage tanks, filters, and pumps in addition to adding municipal water to your rainwater collection system.

The size of your roof and the amount of rainfall you can anticipate in your area are additional considerations. For instance, every inch of rainfall on a 1,000 square foot roof can yield up to 600 gallons of water collection. You can figure out how much water you can collect throughout the year by multiplying the size of your roof by the average annual rainfall in your area.

In conclusion, figuring out your garden's water requirements is crucial to installing a rainwater harvesting system that works.

You can create a system that satisfies your needs and encourages sustainability by taking into account elements like the size of your garden, the kinds of plants you're growing, and the local climate and weather patterns.

Choosing the right system for your garden

Making the best choice for your garden's rainwater harvesting system is crucial because it can affect both the sustainability of your water use and the health of your plants. When choosing a system, it's crucial to take into account aspects like the size of your garden, the climate where you live, and your budget.

Whether you select a straightforward rain barrel or a more intricate underground tank, each system has advantages and disadvantages of its own. You can make an informed choice that will support the health and vitality of your garden while protecting limited water resources by taking the time to research and comprehend your options. Let's see together which are:

Rain Barrels

One of the simplest and most affordable options for rainwater harvesting is a rain barrel. These barrels are typically made of plastic and can hold up to 55 gallons of water. Rain barrels are easy to install and can be placed under a downspout to collect water from your roof. They can be used to water plants, wash cars, and even flush toilets. However, rain barrels may not be sufficient for larger gardens or for areas with long, dry periods.

Pros:

- Inexpensive and easy to install
- Suitable for smaller gardens and tight spaces
- Can be used for a variety of purposes (watering plants, washing cars, etc.)
- Can be connected to a hose for easy distribution of water

Cons:

- Limited capacity (typically around 55 gallons)
- May not be sufficient for larger gardens or areas with long dry periods
- Can be unsightly if not properly concealed

Above-Ground Tanks

Above-ground tanks are larger than rain barrels and can hold up to several thousand gallons of water. They are typically made of plastic, fiberglass, or metal and can be placed on a platform or stand. Above-ground tanks are a good option for larger gardens or for areas with longer dry periods. They can be connected to downspouts to collect rainwater, and some systems may include a pump to distribute the water to different parts of your garden.

Pros:

- Larger capacity (up to several thousand gallons)
- Suitable for larger gardens or areas with longer dry periods
- Can be connected to downspouts for easy collection of rainwater
- Can be fitted with a pump for easy distribution of water

Cons:

- May take up more space than a rain barrel
- Can be more expensive than a rain barrel
- May require a platform or stand for proper installation

Underground Tanks

Underground tanks are a more discreet option for rainwater harvesting. They can hold up to several thousand gallons of water and are typically made of plastic or concrete. These tanks can be connected to downspouts, and a pump may be needed to distribute the water to different parts of your garden. Underground tanks are a good option for those who want to maximize their rainwater collection without taking up space in their yard.

Pros:

- Discreet and out of sight
- Large capacity (up to several thousand gallons)

- Suitable for larger gardens or areas with longer dry periods
- Can be connected to downspouts for easy collection of rainwater
- Can be fitted with a pump for easy distribution of water

Cons:

- More expensive than rain barrels or above-ground tanks
- Installation can be more complex
- May require excavation and disruption of the yard

Green Roofs

Green roofs are a unique type of rainwater harvesting system that involves covering your roof with plants. This system not only collects rainwater but also provides insulation and reduces urban heat island effects. Green roofs can be more expensive to install and maintain than other systems, but they offer numerous benefits for those who are willing to invest in them.

Pros:

- Collects rainwater and provides insulation
- Reduces urban heat island effects
- Improves air quality
- Aesthetic appeal

Cons:

- More expensive to install and maintain than other systems
- Requires specialized expertise and equipment
- May not be suitable for all types of roofs
- Requires regular maintenance and upkeep

Equipment & Supplies Needed

When it comes to setting up a rainwater harvesting system, there are several equipment and supplies that you will need. The specific items you need will depend on the type of system you choose, but here are some common items to consider:

Collection system: To collect rainwater, you will need a collection system such as gutters and downspouts. You may need to install additional downspouts or redirect existing ones to ensure that water flows into your collection system.

Storage tank: A storage tank is needed to hold the collected rainwater. Tanks come in various sizes and materials, such as plastic, metal, or concrete. Consider the capacity needed for your garden's water needs and the space available for the tank.

Filter: Rainwater can contain debris and contaminants, so a filter is needed to remove impurities before the water enters the storage tank. There are several types of filters available, including mesh screens and sediment filters.

Pump: If your rainwater storage tank is located below ground or far from your garden, a pump will be needed to move the water from the tank to your garden. Electric pumps are the most common, but manual pumps are also available.

Hose or irrigation system: To distribute the harvested rainwater to your garden, you will need a hose or irrigation system. Consider the size of your garden and the amount of water needed to determine the appropriate type of irrigation system.

Overflow system: In the event of heavy rainfall, an overflow system is needed to redirect excess water away from your garden and foundation. This can be achieved through the use of a downspout diverter or an overflow pipe.

You may need various tools and equipment to install your rainwater harvesting system:

Ladder: A ladder is needed to access the roof and install gutters and downspouts. Choose a sturdy ladder that can safely support your weight and extend to the appropriate height for your home's roof.

Drill: A drill is necessary to create holes in the gutters and downspouts for mounting brackets and other attachments. Consider a cordless drill for greater flexibility and ease of use.

Saw: A saw may be required to cut gutters and downspouts to the appropriate length. A hacksaw or reciprocating saw is typically used for this task.

Pipe cutters: If you are working with plastic or metal pipes, you will need pipe cutters to cut them to the appropriate length.

Filter wrench: If you are installing a filter, you will need a filter wrench to tighten and remove the filter housing.

Tape measure: A tape measure is essential for measuring and marking the location for mounting brackets, gutters, and downspouts.

Safety gear: It is essential to wear safety gear when installing a rainwater harvesting system. This includes gloves to protect your hands from sharp edges and debris, eye protection to prevent injury from flying particles, and sturdy shoes with good traction to prevent slipping on wet surfaces.

Other supplies: Other supplies that may be needed include screws, mounting brackets, sealant, and Teflon tape.

You can safely and effectively install a rainwater harvesting system and let your garden benefit from it if you have the right tools and equipment. When using these, always abide by the manufacturer's instructions and safety precautions.

CHAPTER 4

Setting Up Your Rainwater Harvesting System

Installing a rainwater harvesting system is an excellent way to save water, reduce utility costs, and reduce your environmental impact. But, with so many different parts and configurations to consider, it can be challenging to know where to start.

We will walk you step-by-step through the installation of your rainwater harvesting system in this chapter. Everything will be covered, including choosing the best system for your requirements, getting your site ready, installing your system, and making sure it is functioning properly. You will have the information and self-assurance required to set up a rainwater harvesting system that is effective, efficient, and catered to your unique needs by the end of this chapter. then let's get going!

How to install a rainwater collection system

Installing a rainwater collection system is an economical and environmentally friendly solution for watering the garden. Near the downspout, connected to it by a pipe, the recovery tank collects and stores rainwater. Here are all the steps to follow to install it properly:

Choosing the right location: Your rainwater collection system's effectiveness depends on where you place it. You should pick a location that is near your garden or landscaping and gets enough light. Avoid areas with bad drainage or a history of flooding. Additionally, you want to pick a flat, sturdy surface that can support the weight of your water-filled tank. Because you will need to be able to access the tank to perform maintenance and cleaning, accessibility should also be taken into account.

Preparing the site: After deciding on the ideal location for your tank, you must prepare the area. Start by clearing away any clutter or obstacles that might prevent your tank from being installed. After that, level the ground to make sure your tank will be balanced and stable. Add a layer of gravel or sand if necessary to aid in drainage. To ensure proper drainage, make sure the site is level and has a small slope away from the tank.

Installing the tank: Your tank can now be installed. Use a diverter or a flexible hose to connect your downspout to the tank's inlet to get started. This will make it possible for rainwater to enter the tank from your gutter. To prevent debris from entering the tank, install a screen filter at the inlet. In order to prevent your tank from overflowing during periods of heavy rainfall, you will also need to install an overflow outlet. Connect the overflow outlet to a dry well or a drainage system.

Installing the distribution system: It's time to set up your distribution system once your tank has been installed. In order to move the water to your garden or landscape, you must connect the outlet of your tank to a pump. Install pipes and hoses to distribute the water where it is required after the pump has been connected. To make the most of the collected water, you can also set up a soaker hose or a drip irrigation system.

Maintenance and troubleshooting: Once your rainwater collection system is operational, it is crucial to perform routine maintenance to make sure it is operating as intended. Look for leaks in your system, and fix any broken parts. To keep debris from entering your system, regularly clean your gutters and downspouts. As advised by the manufacturer, inspect and maintain your pump and other parts. Additionally, you ought to empty and clean your tank once a year to get rid of any accumulated sediment or debris.

Here are all the materials and technical steps involved in installing a rainwater storage system:

Materials:

- Downspout diverter
- PVC pipes
- PVC cement
- Elbow fittings
- Teflon tape
- Hose clamps
- Container (such as a barrel or tank)
- Spigot
- Silicone caulk

- Level
- Saw
- Drill
- Screws
- Shovel

Instructions:

1. Determine the location: Choose a location for your rainwater collection system that is close to the area where you will be using the water. It should also be in an area that receives ample rainfall and has good drainage.

2. Install the downspout diverter: Install a downspout diverter onto the gutter downspout where you want to collect the rainwater. This will divert water from the downspout into the collection container.

3. Connect PVC pipes: Measure and cut PVC pipes to connect the downspout diverter to the collection container. Use elbow fittings to create a bend if necessary. Apply PVC cement to the joints to secure them in place.

4. Add a filter: Install a filter to remove debris and contaminants from the water. Attach a mesh screen or filter to the end of the PVC pipe that leads into the collection container.

5. Install a spigot: Install a spigot near the bottom of the collection container to allow easy access to the water. Use a

hole saw to cut a hole for the spigot, and secure it in place with a hose clamp.

6. Seal the container: Ensure the container has a lid to prevent debris and insects from entering. Use silicone caulk to seal around the lid and spigot to prevent leaks.

7. Test for level: Use a level to ensure the collection container is level. If it is not level, use a shovel to level the ground beneath it.

8. Secure the container: Secure the container in place by attaching it to a stable surface, such as a wall or fence, using screws.

9. Test the system: Test the rainwater collection system by allowing rainwater to flow into it. Check for leaks and ensure the water flows smoothly through the system.

How to Properly Position & Maintain The System

Your rainwater harvesting system must be installed and maintained correctly to operate at its best and last the longest. The slope of the land, the proximity to the garden, the accessibility of the system, and the overall aesthetic must all be considered when choosing the location for your system. A properly positioned system will reduce the need for extra piping and make maintenance easier.

Making sure the gutters and downspouts are clear of obstructions and clean is a crucial component of system maintenance. Debris accumulation can cause overflow or blockages in the pipes, which will lessen the system's efficiency. Regular gutter and downspout cleaning will keep the system clear of clogs and operate efficiently.

In order to avoid contamination and guarantee the quality of the harvested water, the storage tank of the system needs to be properly maintained in addition to being cleaned. For any indications of wear and tear or damage, such as cracks or leaks, the tank should be regularly inspected. Additionally, the tank needs to be cleaned periodically to get rid of any debris or sediment that might build up over time.

Additionally, it's crucial to regularly check the water's quality, particularly if it will be consumed or used to water delicate plants. Water quality testing can help identify any potential issues, such as high contaminant levels or low pH levels, and enable the taking of suitable corrective actions. Finally, it is crucial to adhere to appropriate safety procedures when maintaining and using the system. When handling the system, safety gear like gloves and eye protection should be worn, and all electrical components should be installed and grounded in accordance with local codes and regulations.

You can ensure your rainwater harvesting system operates at peak efficiency and longevity while reaping the advantages of sustainable water use by positioning it correctly and keeping it well-maintained. Below are all the practical steps you need to know to position, maintain and protect your rainwater system, along with all the essential tools.

Positioning Your System

- Determine the best location for your system based on the collection area and intended use for the harvested water.

- Ensure that the area is level and stable to prevent shifting or tipping of the tanks.

- Install the system at least 10 feet away from any septic systems or drain fields to avoid contamination.

- Make sure that gutters and downspouts are properly positioned to channel water into the collection tank.

- If using a pump, make sure it is placed in a location that allows for proper ventilation and accessibility for maintenance.

Maintaining Your System

- Regularly inspect the collection tank for signs of damage or leaks.

- Clean gutters and downspouts at least twice a year to prevent debris from clogging the system.

- Keep the area around the collection tank free of debris to prevent contamination.

- Check for signs of algae growth, especially during warmer months, and add an algaecide if necessary.

- Flush the system at least once a year to remove any accumulated sediment or debris.

- Replace any damaged or worn-out parts as needed to prevent leaks or malfunctions.

Protecting Your System

- Cover the collection tank with a screen or lid to prevent debris, animals, or insects from entering the system.

- Install a first flush diverter to divert the first flush of rainwater, which can contain debris and contaminants, away from the collection tank.

- Use a UV-resistant tank to prevent damage from sunlight exposure.

- Consider adding a backup system, such as a secondary storage tank or a municipal water supply, in case of extended dry periods.

Troubleshooting

Rainwater harvesting systems, like any other system, occasionally encounter problems that call for troubleshooting. The good news is that many of these problems have simple, low-cost solutions that can be applied right away. Following are a few of the most prevalent problems and potential fixes:

- **Clogging of gutters and downspouts:** Leaves, twigs, and other debris can collect in gutters and downspouts, causing them to become clogged and preventing water from flowing into the collection tank. To avoid this, it's important to regularly clean gutters and downspouts. You can use a gutter scoop or a hose to remove debris. You can also install gutter guards or screens to prevent debris from entering the gutters in the first place.

- **Overflowing of the collection tank:** If your collection tank is overflowing, it could be due to heavy rainfall or a blockage in the tank's outlet pipe. One solution is to install an overflow pipe that redirects excess water away from the tank and into another area of the garden. Another solution is to install a float valve or an overflow alarm that alerts you when the tank is full.

- **Algae growth in the collection tank:** Algae can grow in the collection tank if sunlight reaches the water, which can lead to unpleasant odors and clogging of pipes. To prevent this, you can install a cover over the collection tank to block sunlight. You can also add a small amount of bleach or other biocides to the water to prevent the growth of algae.

- **Mosquito breeding in the collection tank:** Standing water in the collection tank can attract mosquitoes, which can be a health hazard. To prevent this, you can install a screen or mesh over the inlet to the tank to prevent mosquitoes from entering. You can also add mosquito dunks or other larvicides to the water to kill mosquito larvae.

- **Freezing of the collection system in cold weather:** In colder climates, rainwater harvesting systems can freeze, which can cause pipes to burst and damage the system. To prevent this, it's important to drain the system before winter and to insulate any exposed pipes. You can also install a heat cable or heat tape to keep the pipes from freezing.

Overall, many of these typical issues can be avoided with routine cleaning and maintenance of the rainwater harvesting system. In order to prevent expensive repairs, it's crucial to keep an eye out for any signs of damage or malfunction and to address any problems as soon as they arise.

CHAPTER 5

DIY Rainwater Harvesting Projects

DIY rainwater harvesting projects might be the ideal choice for you if you want to start collecting rainwater but want to save money or prefer a hands-on approach. These projects can not only be a rewarding and enjoyable way to build your own rainwater collection system, but they can also be tailored to your individual requirements and garden layout.

This chapter will give you helpful advice and step-by-step instructions for making your own DIY rainwater harvesting projects, from straightforward barrel systems to more intricate setups. This chapter will give you the information and motivation to begin creating your own rainwater harvesting system, whether you are an experienced DIY enthusiast or are just getting started.

Building Your Own Barrel

Materials:

- A 55-gallon food-grade barrel
- A spigot
- A bulkhead fitting

- A PVC pipe and elbow joint
- Teflon tape
- Mesh screen
- Silicone sealant
- Tools: drill, saw, wrench, pliers, measuring tape, and marker

Steps:

1. Clean and sanitize the barrel thoroughly, inside and out, with soap and water.
2. Measure and mark the location of the spigot near the bottom of the barrel.
3. Drill a hole for the spigot, using a bit slightly smaller than the diameter of the spigot.
4. Install the bulkhead fitting into the hole, following the manufacturer's instructions, and tighten with a wrench.
5. Wrap Teflon tape around the threads of the spigot and screw it into the bulkhead fitting.
6. Cut a hole near the top of the barrel for the PVC pipe and elbow joint, using a saw or hole saw.
7. Insert the PVC pipe and elbow joint into the hole, using silicone sealant to ensure a tight fit.
8. Cut a piece of mesh screen to fit over the opening of the PVC pipe, and secure it in place with silicone sealant.

9. Position the rain barrel in a suitable location beneath a downspout, and elevate it on cinder blocks or a stand, if necessary.

10. Attach the downspout to the PVC pipe with a flexible elbow joint, and secure it in place with zip ties or metal clamps.

11. Fill the rain barrel with water and check for leaks around the spigot and PVC pipe.

12. Start using your rain barrel to collect and store rainwater for your garden or other outdoor uses.

Building Your Own Cistern

Materials:

- Large plastic container (such as a food-grade drum or water tank)
- Hose spigot
- PVC pipe (1 inch in diameter)
- PVC cement
- Saw
- Drill
- Silicone caulk

Steps:

1. Choose a location for your cistern that is close to a downspout from your roof.

2. Determine the size of the container you will need based on your water needs and the size of your roof. A good rule of thumb is to use a container that can hold at least 50 gallons.

3. Clean the container thoroughly with soap and water, making sure to remove any residue or debris.

4. Cut a hole near the top of the container that is large enough to fit the hose spigot.

5. Install the hose spigot into the hole, making sure it is secure and watertight. Apply silicone caulk around the edges to ensure a tight seal.

6. Cut a hole near the bottom of the container that is large enough to fit the PVC pipe.

7. Install the PVC pipe into the hole using PVC cement. Make sure it is secure and watertight.

8. Cut another hole near the top of the container that is large enough to fit the other end of the PVC pipe.

9. Install the other end of the PVC pipe into the hole, making sure it is secure and watertight.

10. Position the cistern under the downspout from your roof so that the PVC pipe is aligned with the downspout.

11. Attach the PVC pipe to the downspout using a flexible connector or other fitting that will allow for easy removal.

12. Secure the cistern in place and fill it with water.

Building Your Own Underground Storage Tank

Materials:

- 2 plastic water tanks (make sure they fit within the hole you will dig)
- PVC pipes and fittings
- Concrete blocks
- Concrete mix
- Gravel
- Shovel
- Tarp or other waterproofing material
- Tape measure
- Level

Steps:

1. Select a location for your underground storage tank that is level and free of obstructions. It should be close to the area where you plan to use the stored water.

2. Dig a hole that is large enough to accommodate the two water tanks with a space of at least 6 inches between them. The depth of the hole should be determined based on the height of the tanks and how much of them you want to be

buried. Be sure to slope the sides of the hole so that it is wider at the top than at the bottom.

3. Drill holes in the top and bottom of each tank to allow for water flow. Attach the PVC pipes and fittings to the bottom of the tanks using waterproof adhesive or silicone sealant.

4. Lower the tanks into the hole, making sure they are level and spaced apart by at least 6 inches. Place concrete blocks around the tanks to hold them in place and ensure they do not shift.

5. Connect the PVC pipe to the tanks and run it to the area where you plan to use the stored water. Be sure to install a valve or spigot to control the flow of water.

6. Fill the area around the tanks with gravel to help with drainage and prevent the tanks from shifting.

7. Mix the concrete according to the manufacturer's instructions and pour it into the hole, covering the tanks completely. Use a trowel to smooth the surface of the concrete.

8. Lay a tarp or other waterproofing material over the top of the tank before filling the hole. This will help prevent water from seeping into the tank and contaminating the stored water.

9. Backfill the hole with the excavated soil, making sure to compact it into layers. Leave a few inches of space at the top for planting or landscaping.

10. Fill the tanks with water and test the system to ensure proper operation. Monitor the system regularly to ensure that it is functioning correctly and that the stored water is free of contaminants.

Building Your Own Irrigation System

Materials:

- Rain barrel or cistern
- PVC pipe (size will depend on the size of your garden and water needs)
- PVC fittings (elbows, tees, end caps, etc.)
- Drip irrigation tubing
- Drip irrigation emitters
- Hose clamps
- Scissors or pipe cutter
- Hole punch
- Teflon tape
- Valve (optional)

Instructions.

1. Choose the location for your rain barrel or cistern. It should be close to your garden for easy access and to minimize the length of PVC pipe needed.

2. Install a spigot or valve at the bottom of your rain barrel or cistern. This will allow you to control the flow of water.

3. Attach a PVC pipe to the spigot or valve using a PVC fitting. The size of the PVC pipe will depend on the size of your garden and water needs. A larger garden will require a larger PVC pipe.

4. Lay the PVC pipe in your garden, making sure it is sloped slightly downhill to allow for proper drainage. Use PVC fittings to turn corners or make connections.

5. Punch holes in the PVC pipe where you want the drip irrigation emitters to be located. Use a hole punch to make the holes.

6. Insert the drip irrigation emitters into the holes in the PVC pipe.

7. Attach the drip irrigation tubing to the PVC pipe using hose clamps. The tubing should run from the PVC pipe to each plant in your garden.

8. Install the drip irrigation emitters onto the tubing. The number of emitters will depend on the size of your garden and water needs.

9. Turn on the spigot or valve to start the flow of water. Adjust the valve to control the flow of water to the garden.

10. Monitor the garden to ensure that each plant is receiving adequate water. Adjust the placement and number of drip irrigation emitters as needed.

Notes:

- To avoid leaks, use Teflon tape on all PVC fittings.
- A valve can be installed in the PVC pipe to allow you to turn off the water flow to the garden when it is not needed.
- A filter can be added to the system to remove debris and improve water quality.

How to Setup The Perfect Rain Garden

Materials:

- Shovel
- Garden hose
- Mulch
- Compost or organic fertilizer
- Rain barrel or other rainwater harvesting system
- Drip irrigation system or watering can

Steps:

1. Choose the location: Pick an area in your garden that gets plenty of sunlight and has good drainage. It's also important

to choose an area that's close to your rainwater harvesting system.

2. Prepare the area: Clear any weeds or debris from the area and use the shovel to dig the soil to a depth of at least six inches. This will help to loosen the soil and improve drainage.

3. Add organic matter: Add a layer of compost or organic fertilizer to the soil and mix it in using the shovel. This will help to improve soil health and provide nutrients to your plants.

4. Install a rainwater harvesting system: Install a rain barrel or other rainwater harvesting system near your garden. Connect it to your gutters or use a garden hose to collect rainwater.

5. Install a drip irrigation system or use a watering can: Drip irrigation systems are a great way to water your plants efficiently and minimize water waste. If you're on a tight budget, you can also use a watering can to water your plants by hand.

6. Choose the right plants: Select plants that are well-suited for your region and soil type. Native plants are often a good choice as they are well-adapted to the local environment and require less maintenance.

7. Mulch: Apply a layer of mulch around your plants to help retain moisture in the soil and prevent weeds from growing.

Mulch can also help to regulate soil temperature and protect your plants from extreme weather conditions.

8. Water regularly: Water your plants regularly using rainwater collected from your rainwater harvesting system. If you notice that your plants are getting too much or too little water, adjust your watering schedule accordingly.

9. Maintain your garden: Regularly check your garden for pests, diseases, and other issues. Remove any dead or diseased plants and prune your plants as needed to promote healthy growth.

CHAPTER 6

Using Harvested Rainwater in Your Garden

Imagine admiring your flourishing plants and vegetables while strolling through your garden, knowing that you were a significant contributor to their development. You can create a sustainable and eco-friendly solution that benefits your plants and the environment by collecting and using harvested rainwater in your garden.

This chapter will discuss different ways to use collected rainwater in your garden and how doing so can make you a more responsible and effective gardener. There are countless options, from feeding your soil to watering your plants. Let's explore the advantages of using harvested rainwater in your garden now.

Best practices for using harvested rainwater

The environment and the health of your plants can both benefit from using harvested rainwater in your garden. To get the most out of it, you should adhere to some best practices. To make sure you get the most out of your rainwater harvesting system, we will examine the dos and don'ts of using harvested rainwater in your garden in this chapter.

The most important thing to remember is to put your collected rainwater in an appropriate container. By doing this, you can ensure that the water is safe for your plants and that it is not contaminated. Use food-grade containers made of plastic, fiberglass, or stainless steel, it is advised. Use of untreated metal or PVC containers, as well as other materials that can release harmful chemicals into water, should be avoided.

It's important to think about the water's quality before using harvested rainwater to water your plants. The water from your rainwater harvesting system should be suitable for your plants if it has been installed and maintained correctly. Testing the pH and nutrient levels of the water on a regular basis is still a good idea. Here is a step-by-step process on how to test water pH quickly and easily:

Materials needed:

- pH test strips or a pH meter
- A clean container to collect water sample
- Distilled or deionized water (if using pH meter)

Steps:

1. Collect a clean water sample from your rainwater harvesting system. Use a clean container and avoid touching the inside of the container with your fingers.
2. If using pH test strips, dip the strip into the water sample and wait for the specified amount of time according to the instructions.

3. If using a pH meter, fill the meter's electrode with distilled or deionized water and turn on the meter. Once the meter has stabilized, dip the electrode into the water sample and wait for the reading to stabilize.

4. Read the pH value from the test strip or pH meter display. The ideal pH for most plants is between 6.0 and 7.0, but it can vary depending on the specific plant species.

5. If the pH value is outside of the ideal range, adjust the pH level accordingly. To lower the pH, add an acidic substance like vinegar or citric acid. To raise the pH, add a basic substance like baking soda or agricultural lime.

6. Retest the pH after adjusting the level to ensure it is within the desired range.

It's crucial to make good use of the rainwater you collect. To reduce water loss due to evaporation, one best practice is to water your plants in the early morning or late at night when it is cooler outside. Additionally, you should avoid overwatering your plants because it can result in root rot and other problems.

Finally, it's critical to conserve rainwater by using it properly. To minimize runoff and deliver water directly to your plants' roots, think about using drip irrigation or a soaker hose. Mulch can also help the soil retain moisture, requiring less watering overall. By adhering to these recommendations, you can lessen your impact on the environment while making sure that the rainwater you collect for your plants is secure and efficient.

How to Incorporate Rainwater Harvesting Into Your Gardening Practices

As a gardener, you have a precious resource right at your fingertips: harvested rainwater. Not only is this water free, but it is also beneficial to your plants and reduces your reliance on expensive municipal water. *But what are the best ways to use your collected rainwater in your garden?* Let's explore some of the most common and effective methods.

Optimizing the use of harvested rainwater is key to ensuring its effectiveness. Gardeners can use rainwater in various ways, including irrigation, watering plants, and washing outdoor furniture or tools. To ensure optimal usage, consider using a drip irrigation system, which can deliver water directly to the roots of your plants and minimize water loss due to evaporation. You can also use a rain gauge to track rainfall levels and adjust your watering schedule accordingly.

- **Irrigation:** Irrigation is one of the most popular uses of rainwater collection. To water your plants slowly over time, you can set up a simple drip irrigation system with hoses, pipes, and emitters. In addition to conserving water, this lowers the chance of waterlogging or overwatering, both of which can harm your plants.

- **Watering plants:** Using a bucket or watering can, you can use the rainwater you've collected to water your plants directly. This enables you to target particular plants, such as those in pots or hanging baskets, that may require more water than others.

- **Washing outdoor furniture or tools:** You can save money on cleaning supplies and lessen your impact on the environment by using rainwater to clean outdoor furniture or tools.

- **Composting**: Adding rainwater to your compost pile is another way to use it in your garden. This can help your compost retain more moisture and hasten the decomposition process.

- **Mixing with fertilizers**: You can also combine the rainwater you've collected with organic fertilizers to make a nutrient-dense solution you can use to feed your plants. This can be done with the help of a straightforward mixing container and offer a cost-effective and environmentally friendly substitute for store-bought fertilizers.

Finally, incorporating rainwater harvesting into your gardening practices can lead to significant cost savings and environmental benefits. For example, a small rain barrel can collect up to 50 gallons of water during a single rainfall, reducing your reliance on municipal or well water. Additionally, using harvested rainwater can reduce the amount of runoff that contributes to water pollution and can help promote a healthy garden ecosystem by providing plants with natural and nutrient-rich water.

Tips for maximizing the benefits of harvested rainwater for your garden

You are aware as a gardener that water is a valuable resource for your plants. But did you know that you can access a plentiful and free source of water right now? Harvesting rainwater is an excellent way to reduce your water bill and give your garden the hydration it needs. Furthermore, rainwater is devoid of the additives present in tap water, which can promote the health and growth of your plants. *So why not make use of this fantastic resource and provide your garden with the nutrition it needs?* Here are a set of smart tips for maximizing the benefits of harvested rainwater for your garden:

- **Choosing the right plants:** It's important to choose plants that are suited to your climate and soil conditions. Drought-tolerant plants are great for gardens that rely on rainwater harvesting because they require less water. Some examples of these plants include succulents, cacti, and native plants. These plants have adapted to survive in arid environments and can help you save water while still creating a beautiful garden.

- **Using mulch:** Mulching your garden beds with organic materials like straw, leaves, or wood chips can help retain moisture in the soil and reduce water evaporation. This means you won't have to water your plants as often, and your plants will receive more of the rainwater you collect. Additionally, mulch can help control weeds, reduce soil erosion, and improve soil health.

- **Consider a drip irrigation system:** Drip irrigation systems minimize water loss from evaporation or runoff by delivering water directly to the base of your plants. This makes them a great option for gardens that rely on rainwater harvesting because they maximize the efficiency of the water you have collected. Drip irrigation systems are also easy to install and can be customized to fit the specific needs of your garden.

- **Install a rain sensor:** A rain sensor is a device that automatically turns off your irrigation system when it senses rain. This prevents overwatering and ensures that your plants receive only the water they need. A rain sensor can also help you save water and reduce your water bill.

- **Collect and use gray water:** Gray water is the wastewater generated from activities such as bathing or doing laundry. This water can be collected and used to water your plants. While it is not safe for drinking, it can be a great source of water for your garden. Make sure to use eco-friendly soaps and detergents to ensure that the gray water is safe for your plants.

- **Use a rain barrel with a spigot:** A rain barrel with a spigot allows you to easily access the water you have collected for use in your garden. It also makes it easy to divert excess water to a different area, such as a nearby lawn or garden. Rain barrels come in many sizes and styles, so you can choose the one that best fits your needs and budget.

- **Clean your gutters and downspouts regularly**: Clean gutters and downspouts ensure that water flows freely and prevents debris from entering your rainwater collection system. This helps to maintain the quality of the water you have collected. Regular cleaning also helps to prevent damage to your gutters and downspouts, which can be expensive to repair or replace.

In conclusion, using collected rainwater in your garden is not only a wise and environmentally friendly choice, but it can also make gardening more enjoyable and fulfilling. You can feed your plants and encourage a healthier environment by using nature's power. Therefore, invest the time in installing a rainwater collection system and enjoy the benefits for many years to come. You'll not only save money and resources, but you'll also improve the health of the environment and the health of your garden.

CHAPTER 7

Choosing the Right Plants for Your Rainwater Garden

The plants you choose to grow in your garden must be carefully chosen if you want them to benefit from rainwater harvesting. When it comes to water requirements and toleration, not all plants are created equal. While some plants can survive periods of drought or heavy rain, others need more frequent watering.

You can create a sustainable, low-maintenance garden that looks lovely and aids in water conservation by picking the right plants for your rainwater garden. In this chapter, we'll cover how to design a rain garden to maximize water usage and plant health, the significance of choosing plants that are suitable for rainwater irrigation, how to maintain your rain garden to ensure its long-term success, and more.

Well suited

Are you trying to find plants for your rainwater garden that will thrive? The secret to creating a stunning, long-lasting landscape that can withstand changing weather conditions is selecting the appropriate plants for your garden. Your water usage can be decreased, you can save money on your water bill, and you can encourage a healthy ecosystem in your garden by choosing plants that do well with rainwater irrigation. We will examine the qualities of plants that can survive in a rainwater garden in this chapter, along with helpful advice on how to select and take care of them. You can design a stunning garden that is both beautiful and environmentally friendly with the right plants.

- **Native Plants:** Because they have adapted to the local climate and soil characteristics, native plants are excellent candidates for irrigation with rainwater. They are frequently more pest and disease resistant and require less water than non-native plants. Consider native plants' growth patterns, needs for sun and shade, and blooming periods when making your selection. Native plants, like California poppy, yarrow, and sage, are excellent options for rainwater gardens in California, for instance.

- **Perennial Plants:** Due to their resiliency and lower maintenance needs than annual plants, perennial plants are a fantastic option for rainwater gardens. They also grow roots that are deeper, which enables them to reach water that is stored in the soil at a deeper level. Take into account the perennial plants' growth patterns, needs for sun and shade, and blooming periods. In the Midwest, for instance, perennials like phlox, black-eyed Susan, and coneflower make excellent choices for rainwater gardens.

- **Xeriscaping Plants:** Because they have adapted to dry environments and need little water to survive, xeriscaping plants are ideal for rainwater irrigation. They come in a variety of colors and textures and are frequently low maintenance. Take into account the xeriscaping plants' growth patterns, needs for sun and shade, and blooming periods. For instance, xeriscaping plants like agave, yucca, and desert marigold are excellent options for rainwater gardens in Arizona.

- **Edible Plants:** Because they provide fresh produce and require less irrigation than conventional gardens, edible plants are a great option for rainwater gardens. When selecting edible plants, take into account their growth patterns, needs for sun and shade, and harvest times. In the Pacific Northwest, for instance, edible plants like lettuce, kale, and chard make excellent choices for rainwater gardens.

- **Water-Loving Plants:** Water-loving plants are a great option for rainwater gardens because they can help absorb extra water and thrive in moist environments. Consider a plant's growth habits, light and shade requirements, and blooming period when selecting a water-loving species. For instance, water-loving plants like cardinal flower, joe-pye weed, and swamp milkweed are excellent options for rainwater gardens in the Southeast.

Overall, it is important to choose a mix of plants that are well-suited for your local climate and soil conditions, as well as your personal preferences. Consider the overall design of your rainwater garden and choose plants that will work well together in terms of height, color, and texture. It is also important to consider the maintenance requirements of each plant and choose plants that fit within your schedule and abilities.

Designing a Rain Garden To Maximize Water Usage And Plant Health

A successful and sustainable garden that utilizes rainwater harvesting requires the design of a rain garden. A rain garden's main function is to gather and store rainwater in the soil so that plants can gradually absorb it. The location, scale, and water needs of the plants should all be taken into account when designing a rain garden.

Location and Site Preparation

Select a location for your rain garden that receives enough runoff from driveways, impervious surfaces, and rain. Additionally, the location should be far from any structures and septic tanks. Examine the soil before starting the garden and make any necessary adjustments to ensure good drainage. To enhance the quality and drainage of the soil, if necessary, add organic matter. To make sure the plants you select are suitable for the soil conditions, test the soil to determine the pH and nutrient levels.

Example: You might need to amend the soil with sand or other materials if your area has heavy clay soils in order to improve drainage. As an alternative, you might need to add organic matter to your site's sandy soil to improve water retention.

Shape and Size

Determine the size and shape of your rain garden based on the amount of runoff it will receive. Generally, the size of the garden should be one-third the size of the impervious surface that it will be collecting runoff from. The shape of the garden should be designed to encourage water to flow towards the center, where it will be absorbed by the plants.

Example: The ideal size for a rain garden on a 1,000 square foot roof would be about 300 square feet. The form could be kidney-shaped, round, oval, or any other shape with a small depression in the middle to promote water flow.

Plant Selection

Pick plants that can survive in both wet and dry conditions and are native to your area. Deep-rooted plants are especially good at absorbing and filtering water. Choose a range of plants that bloom at various times of the year to provide interest all year long and draw pollinators.

Example: Due to their tolerance for damp conditions, native plants found in the Southeastern United States like swamp milkweed, blue flag iris, and cardinal flower make excellent choices for rain gardens.

Mulch and Maintenance

Mulch the garden to help weeds grow less and retain moisture. To prevent introducing pollutants into the garden, use a natural mulch like wood chips or chopped leaves. Maintain the garden by clearing away trash and dead plants, and check the drainage occasionally to make sure it is working properly.

Example: Using native red alder chips as mulch in the Pacific Northwest can give plants a natural source of nitrogen and aid in moisture retention.

Irrigation and Overflow

Consider using harvested rainwater or other sustainable sources to irrigate your rain garden if the amount of water falling on it is insufficient. Create an overflow outlet to send extra water somewhere else if the garden overflows during a heavy downpour.

Example: Utilizing collected rainwater or greywater in a rain garden can help in arid areas like Arizona by ensuring that the plants receive an adequate supply of water. Flooding during periods of heavy rain can also be avoided by establishing a rain chain or sending excess water into a dry well.

Tips for Maintaining Your Rain Garden

For a rain garden to be successful over the long term, it must be properly maintained after it has been designed and installed. Regular upkeep aids in preserving the garden's aesthetic appeal and usefulness, as well as helping to stop erosion and manage weed growth. This chapter will offer helpful advice on how to maintain your rain garden, including how to keep an eye on water flow, clear away debris, manage weeds, and deal with pests and diseases. We will also go over how to perform seasonal maintenance chores like fertilizing and pruning, as well as how to get your garden ready for harsh weather. You can make sure that your rain garden thrives and offers a host of advantages for both your garden and the environment by paying attention to these suggestions.

- **Regularly check and clean your rain garden:** For rain gardens to continue working properly, regular maintenance is necessary. Regularly inspect your rain garden for any particles, sediment, or other substances that might be impeding water flow. In the garden, take out any dead plants and any leaves that may have fallen. Your rain garden will function more effectively if you clean it frequently.

- **Mulching:** This is a useful technique for keeping your rain garden healthy. Mulch aids in controlling soil temperature, weed suppression, and moisture retention in the soil. Around your plants, add a layer of organic mulch to help the soil retain moisture.

- **Prune and trim your plants:** To keep your rain garden plants healthy and beautiful, regular pruning and trimming are required. When your plants are dormant, prune them to promote new growth and get rid of any dead branches. Your plants will be able to have a strong structure that is less vulnerable to illness and insect infestations as a result.

- **Fertilize and feed your plants**: To thrive, plants in rain gardens need the right kind of nutrition. In the spring, fertilize your garden with compost or a balanced fertilizer to help your plants grow. Follow the manufacturer's recommendations for the correct application rates.

- **Water as needed:** Even though rain gardens are made to catch and store rainwater, you might still need to water your plants when the weather is dry. Keep an eye out for signs of stress in your plants, like wilting, and give them the water they require to stay healthy and grow.

- **Monitor for pests and diseases:** Rain gardens are subject to pests and diseases just like any other type of garden. Check your plants frequently for indications of damage or infestation. The spread of diseases and pests can be halted with early detection and treatment.

- **Manage invasive plants:** Native plants can quickly lose out to invasive ones in your rain garden. Find out how to recognize and control invasive species in your garden.

Invasive plants should be eliminated as soon as you can to stop them from proliferating.

Expert tips:

- If your rain garden is close to trees, be sure to check it frequently for any debris that may fall from the trees and clog the garden and restrict the flow of water.

- To promote new growth and keep your rain garden plants looking neat and healthy, prune them. For instance, to encourage new growth and a fuller, more appealing plant, cut back a butterfly bush in your rain garden by one-third in the spring.

- Your plants may need compost or fertilizer if you notice that they are not growing as well as they should. For instance, if the plants in your rain garden have been there for a while and seem to be having trouble, try adding a layer of compost around them to give them the nutrients they require to thrive.

Water your rain garden as necessary during dry spells to maintain the health of the plants. Give your plants a deep watering, for instance, if you see that they are wilting, to aid in their recovery.

CHAPTER 8

Rainwater Harvesting for Urban Gardening

As interest in sustainable living and the need for fresh produce has grown, urban gardening has become increasingly popular. Urban gardening, however, presents special difficulties, particularly in terms of water management. Access to water sources and a lack of available space can make it difficult to keep a healthy garden. Rainwater harvesting can help in this situation.

Urban gardeners can collect, store, and use rainwater to keep their plants healthy and thriving while reducing their reliance on conventional water sources by using the right tools and techniques. In this chapter, we'll examine the advantages of rainwater collection for urban gardening and offer helpful advice on how to begin using this environmentally friendly technique.

All the Benefits

In addition to aiding in water conservation, rainwater harvesting offers a viable and affordable replacement for conventional watering techniques. We will examine the numerous advantages of rainwater collection for urban gardening, including decreased water costs, enhanced plant health, and minimal environmental impact.

Then, we'll offer helpful advice and examples of how urban gardeners can use rainwater collection to establish thriving, long-lasting gardens in even the most difficult urban settings. Here is a list of specific benefits of rainwater harvesting for urban gardening:

- **Reduced Water Costs:** Rainwater harvesting can help reduce water costs in urban gardening. For example, in drought-prone regions like California, homeowners can save significant amounts on their water bills by using harvested rainwater to irrigate their urban gardens instead of relying solely on municipal water supplies.

- **Improved Plant Health:** Rainwater is free of chlorine and other chemicals commonly found in municipal water supplies, making it an ideal choice for irrigation. Using harvested rainwater can help promote healthier plant growth, as the lack of chemicals and minerals can reduce the buildup of salts in the soil, which can be harmful to plants.

- **Environmental Benefits:** By harvesting and using rainwater for urban gardening, you can reduce your environmental impact by conserving water and reducing the demand on municipal water supplies. This can help reduce the strain on local water sources, which can be especially important in urban areas where water resources may be limited.

- **Enhanced Soil Quality:** Rainwater is typically soft and slightly acidic, which can help enhance soil quality by reducing alkalinity and increasing nutrient availability. This can be especially important in urban areas where soil quality may be poor due to pollution and other environmental factors.

- **Sustainable Gardening:** By using harvested rainwater, you can practice sustainable gardening and reduce your reliance on fossil fuels used to power water treatment and distribution systems. This can help reduce your carbon footprint and promote a more sustainable lifestyle.

Let's imagine an urban gardening scenario in Los Angeles, California, to give some actual examples. Homeowners may have to comply with strict water restrictions and pay high water bills due to the region's frequent droughts. Homeowners can collect and use rainwater for their gardens by installing a rainwater harvesting system, such as a basic rain barrel or a more complex cistern. This can assist them in adhering to water restrictions, help them save money on water, and encourage the growth of healthier plants.

In addition, urban gardeners in Los Angeles can lessen the strain on regional water sources like the Colorado River by substituting harvested rainwater for municipal water supplies. This could encourage a sustainable lifestyle and lessen their impact on the environment.

Finally, urban gardeners can improve the quality of the soil by using rainwater that has been collected. This is crucial in places like Los Angeles where the soil may be polluted by traffic and other sources. Rainwater's slightly acidic composition can aid in lowering soil alkalinity and boosting nutrient availability, which will result in healthier plants and a more vibrant urban garden.

Overcoming Challenges

Urban gardening presents particular difficulties because of the scarce availability of water, poor soil quality, and available space. Although installing a rainwater harvesting system in an urban garden can be a practical solution to these problems, there are a few things to keep in mind.

Limited space for rainwater storage may be a problem for urban gardeners. It might be necessary to use smaller rain barrels or even put in a vertical rainwater harvesting system to make the most of the available space. Vertical systems can be tailored to meet particular needs and attached to walls or fences.

The quality of the soil in urban areas, which is frequently contaminated or deficient in nutrients, is another difficulty. By giving the garden access to clean water, rainwater harvesting can help to alleviate this problem. The type of roofing material and any possible contaminants that might be present in the collected rainwater should both be taken into account. You can make sure the water is suitable for plants by using a filter or another type of treatment.

Urban gardening can also be difficult for those without access to water sources outside or who live in apartments. A portable rainwater harvesting system, like a collapsible rain barrel, may be an efficient solution in this situation. Because of how easily these systems can be moved and stored indoors, it is also possible to use the collected rainwater for indoor plants. It's crucial to take into account any local ordinances and zoning rules that could affect rainwater collection in urban areas. Some cities may have specific rules regarding the use of harvested rainwater or may require permits for rainwater harvesting systems.

Overall, with some creativity and problem-solving, rainwater harvesting can be a valuable tool for overcoming challenges in urban gardening. Urban gardeners can maximize their available space and resources by using smaller rain barrels, setting up vertical systems, thinking about the quality of the soil, and addressing access to water.

Innovative Ways To Use Rainwater In Urban Gardens

Rainwater harvesting is not only a sustainable practice, but it also efficiently lowers water bills and conserves water. Finding creative ways to use rainwater in your garden can help you make the most of this priceless resource, in addition to collecting and storing it. This chapter will explore some creative uses for rainwater in urban gardens and will offer real-world scenarios and useful examples.

- **Drip irrigation:** Because it delivers water directly to the roots of your plants, drip irrigation is a very effective method of watering your garden. Up to 50% less water can be used for irrigation when using drip irrigation instead of conventional irrigation techniques. Using a rain barrel or cistern, drip irrigation systems are simple to set up and can be tailored to your garden's individual requirements. For instance, you can install a drip irrigation system if you have raised garden beds to deliver water to each bed at a specific rate based on the requirements of the plants in that bed.

- **Vertical gardening:** is a well-liked technique for growing plants in constrained urban areas, and when combined with rainwater collection, it can be even more productive. You can increase your growing area while also saving water by utilizing a vertical gardening system. Installing a rain gutter garden, where plants are grown in gutters that are installed on a vertical surface, such as a wall, is one creative way to use rainwater in vertical gardening. You can water your plants without using any municipal water by using rainwater collected in a cistern or rain barrel.

- **Hydroponics**: Using a nutrient-rich solution to feed the plants, hydroponics is a soil-free method of growing plants. Even though hydroponics already uses very little water to grow plants, using rainwater can increase its sustainability.

You can give your hydroponic plants a reliable source of water that is free of contaminants like chlorine or fluoride, which can be bad for plants, by collecting and storing rainwater.

- **Composting:** As a rich source of nutrients for plants, composting is a crucial gardening practice for anyone. In addition to saving water, using rainwater to hydrate your compost pile can hasten the decomposition process. You can easily water your compost pile without using any city water by collecting rainwater in a rain barrel or cistern.

- **Water Features:** Fountains and ponds are two examples of water features that can beautify and calm your urban garden. By using rainwater to fill your water feature, you can conserve water while also creating an attractive focal point in your garden. Utilizing rainwater can also assist in avoiding the accumulation of minerals and other contaminants in your water feature, which can be detrimental to fish and plant life.

You can make the most of this priceless resource by finding creative ways to use rainwater in urban gardens. You can save water, lower your water bills, and make your garden more sustainable by using rainwater for drip irrigation, vertical gardening, hydroponics, composting, and water features. There are countless other ways to utilize this priceless resource in your garden; these are just a few examples of how rainwater harvesting can be incorporated into your urban gardening practices.

How to Setup Your Rainwater Harvesting System

Although installing a rainwater harvesting system for urban gardening may seem difficult, with the right knowledge, it can be a straightforward and worthwhile project. In this chapter, we will guide you through the step-by-step process of setting up your own rainwater harvesting system for urban gardening. Everything from selecting the ideal system for your requirements to installing and maintaining it will be covered. You will be prepared to install your own rainwater harvesting system and take advantage of using rainwater for your urban gardening needs by the end of this chapter.

Step 1: Determining Your Needs for Urban Gardening

Identifying your unique needs is the first step in setting up a rainwater harvesting system for urban gardening. Take into account your urban garden's size, the plants you'll be growing, and the local rainfall. Establish how much water your garden will require and how frequently you'll need to water your plants.

Step 2: Choosing the Best System for Your Urban Gardening Needs

Choose the best rainwater harvesting system for your needs after evaluating your needs for urban gardening. Take into account the size of your garden and the local rainfall. For urban gardening, a variety of rainwater harvesting systems, including barrels, cisterns, and tanks, are available. Find the system that best suits your needs and financial situation.

Step 3: Acquiring the Required Supplies and Equipment

Get the tools and supplies you need after deciding on your rainwater harvesting system for urban gardening. This includes any additional installation-related supplies, such as hoses, connectors, downspout diverters, and tools. Before starting the installation process, make sure you have all the necessary tools and supplies.

Step 4: Installation of Your Rainwater Harvesting System for Urban Gardening

Depending on the type of system you have selected, the installation procedure for your rainwater harvesting system for urban gardening may vary. In general, the procedure entails attaching the hoses to the diverter, connecting the hoses to the rainwater collection system, and connecting the downspout diverter to the gutter system. If necessary, seek professional assistance or carefully adhere to the manufacturer's installation instructions.

Step 5: Setting Up and Maintaining Your Urban Gardening System

For maximum effectiveness, the placement of your rainwater harvesting system for urban gardening is essential. Ensure that it is positioned in an area that gets a lot of rain and is convenient for maintenance. To guarantee that your system is operating properly, routine maintenance is also crucial. Check for leaks or damage to the hoses and clean out any debris from the gutters and downspouts.

Step 6: Using Your Harvested Rainwater in Your Urban Garden

You can begin using the rainwater collected in your garden once your rainwater harvesting system for urban gardening is installed and maintained. Use the rainwater you've collected to water your plants, and think about using it for other gardening tasks like pond or fountain filling. You can lessen your reliance on municipal water supplies and help create a more sustainable urban environment by using harvested rainwater.

Here's an example based on an urban gardening system:

Maggie maintains an assortment of herbs and vegetables on her small balcony in her apartment in a busy city. She made the decision to install a rainwater harvesting system for her urban garden in order to lessen her reliance on tap water and her environmental impact.

Maggie made the decision to buy a small barrel that would fit on her balcony after researching various systems. She gathered the necessary equipment, including a downspout diverter, hoses, and connectors. She carefully followed the manufacturer's instructions and put the system in a balcony corner that got plenty of rain.

Maggie was ecstatic with the outcomes. She noticed that using rainwater increased the health and strength of her plants in addition to reducing her reliance on tap water. In order to produce her own nutrient-rich soil for her garden, she even started a small composting bin.

Her rainwater harvesting system was easy to maintain. Maggie regularly checked the hoses for leaks and cleared any debris from the diverter for the gutter and downspout. To make sure the water was clean, she also added a filter to the system.

Overall, Maggie found installing a rainwater harvesting system to be a simple and rewarding project that helped her meet her sustainability objectives for her urban garden.

CHAPTER 9

Rainwater Harvesting in Different Climates

As we have seen, the practice of collecting rainwater has been used by civilizations all over the world for thousands of years. However, depending on the climate and geographical location, different rainwater harvesting techniques may be used. There are methods and systems that can be tailored to your particular climate, whether you reside in a region with plentiful rainfall or a desert region with limited water resources.

The various factors and options for rainwater harvesting in various climates will be covered in this chapter. We will go over the best methods and procedures for maximizing rainwater harvesting in all climates, from temperate areas to dry deserts.

The importance of climate

The quantity and quality of the water harvested may vary depending on the climate, which also affects evaporation rates and soil types. Farmers and homeowners can implement rainwater harvesting systems that are customized to their unique needs by understanding the unique challenges and opportunities presented by various climates. Even in the most difficult climates, rainwater harvesting can be a dependable and sustainable source of water with the right methods and tools.

Type 1: Arid and Semi-Arid Climates

Rainfall is sporadic and infrequent in arid and semi-arid climates. Because of this, collecting rainwater may be difficult but not impossible. Utilizing methods like contour farming and irrigation that uses the least amount of water is one way to maximize the use of the water collected. Rainwater can be stored in tanks or cisterns for later use, with any extra water being used to recharge the groundwater table. Shading the water storage area and covering the surface with floating balls or other materials can help reduce water loss in areas with high evaporation rates.

Type 2: Temperate and Humid Climates

Arid and semi-arid climates experience less rainfall than temperate and humid climates. Because there is more water available for collection, rainwater harvesting becomes a more practical option. But too much rain can also be a problem because it can cause flooding and soil erosion. Some methods used to control excess rainfall and avoid flooding include rain gardens, green roofs, and permeable pavement. Additionally, during dry spells, irrigation can be augmented by rainwater harvesting.

Type 3: Cold and Snowy Climates

Rainwater harvesting is not always possible in cold and snowy climates during the winter months when snow and ice cover the ground. Rainwater, on the other hand, can be collected in barrels, tanks, or underground cisterns for later use during the spring thaw. In areas where snowfall is common, rainwater harvesting can be combined with snowmelt harvesting, which collects and stores snow for later use. This can be accomplished through the use of simple techniques such as snow fences or more sophisticated systems such as snow collectors.

Rainwater Harvesting in Arid Climates

In arid climates, where rainfall is often scarce and unpredictable, selecting the right rainwater harvesting system is critical. When selecting a system, several factors must be considered, including the size of your property, the amount of rainfall, and the type of crops or livestock you have. Rooftop catchment systems, subsurface dams, and earthworks are examples of common harvesting systems for arid climates.

It is critical to maximize the use of collected rainwater once you have selected the appropriate rainwater harvesting system. This can be accomplished by using proper irrigation techniques, such as drip irrigation or underground irrigation, which reduce water loss due to evaporation. It is also critical to properly store collected rainwater to avoid evaporation and contamination.

Using rainwater for livestock can also help to reduce the demand for costly, imported water sources.

In arid climates, there are several smart tips that can help you make the most of your rainwater harvesting system. Planting drought-resistant crops that require less water, for example, can help to reduce your water demand. It is also critical to maintain your harvesting system on a regular basis to ensure proper operation and to prevent water loss. Taking advantage of any rain events, no matter how minor, can also help to maximize the benefits of your system.

Consider a farmer in Arizona who relies on rainwater harvesting to irrigate his crops. Because of the area's low rainfall, the farmer installed a rooftop catchment system to collect rainwater from his barn and greenhouse. He then waters his crops with a drip irrigation system, which reduces water loss due to evaporation. Furthermore, the farmer grows drought-resistant crops like beans and peppers, which require less water. The farmer can maximize the benefits of his rainwater harvesting system and reduce his reliance on expensive, imported water sources by taking advantage of every rain event.

Rainwater Harvesting in Rainy Climates

Rainy climates provide a unique opportunity for rainwater harvesting due to the abundance of precipitation. However, in order to make the most of the rainy season, a proper system must be in place. We will provide practical solutions and smart tips on rainwater harvesting in rainy climates in the following lines.

Understanding rainfall patterns in your area is the first step in maximizing rainwater harvesting in rainy climates. Investigate the average rainfall in your area and the time of year when it is most likely to rain. This data can assist you in determining the best time to install your rainwater harvesting system as well as when to expect the most precipitation.

The best rainwater harvesting system for your area will be determined by your specific needs and the amount of rainfall expected. To capture and store excess water in rainy climates, a larger capacity system may be required. Furthermore, the quality of the collected water must be considered, as heavy rainfall can sometimes result in debris or contaminants being washed into the collection system. Choosing a system with appropriate filtration and maintenance features can help ensure the quality of the water you collect.

In a rainy climate, designing your rainwater harvesting system requires careful consideration of where and how to collect the water. One option is to channel the water into a collection tank using a flat roof or a sloped roof with gutters. A rain garden or swale is another option for capturing and storing excess water in the soil. It is critical to locate your system in an area that receives adequate rainfall and is easily accessible for maintenance.

Regular maintenance is required to ensure your rainwater harvesting system's longevity and efficiency. In rainy weather, it's critical to clean out any debris that has accumulated in your gutters or collection tank. Using collected rainwater for gardening or other non-potable purposes can also help conserve municipal water supplies and lower your water bills. It is important to note, however, that rainwater should not be used for drinking or cooking unless properly treated.

Rainwater harvesting is ideal in Seattle, which is known for its rainy weather. Rain gardens and cisterns have been installed by the city to collect and store excess rainwater for later use. The Bullitt Center in Seattle, for example, has a 56,000-gallon cistern that collects rainwater from the roof and stores it for use in the building's plumbing and irrigation systems. Furthermore, the city of Seattle provides rebates to residents who install rain gardens or cisterns, thereby encouraging the use of rainwater harvesting systems.

Adapting rainwater harvesting techniques to different climate conditions

The effectiveness of the different rainwater harvesting techniques can vary depending on the climate conditions of the area. Adapting rainwater harvesting techniques to different climates is crucial for maximizing their benefits. We will explore the different adaptations and smart advice that can be done to rainwater harvesting techniques in various climate conditions.

Rainwater Harvesting in Arid Climates

In arid climates, water scarcity is a common issue. To maximize the benefits of rainwater harvesting in such conditions, the following adaptations can be made:

- Use of efficient rainwater catchment systems such as low-tech systems like swales or high-tech systems like rooftop catchment systems with efficient filtration systems.

- Storing the harvested water in underground cisterns to minimize evaporation and keep the water cool, preventing bacterial growth.

- Incorporating drought-tolerant plants in landscaping to minimize water usage.

- Utilizing greywater systems to further reduce water usage.

Rainwater Harvesting in Humid and Wet Climates

In humid and wet climates, the following adaptations can be made to maximize the benefits of rainwater harvesting:

- Use of large storage tanks to accommodate the higher volume of water.

- Utilizing overflow systems to prevent damage to the harvesting system during heavy rainfall.

- Incorporating rain gardens or green roofs to absorb excess rainwater and prevent flooding.

- Utilizing rain barrels or smaller storage tanks for garden irrigation.

Rainwater Harvesting in Temperate Climates

Temperate climates have moderate rainfall patterns, making them suitable for a variety of rainwater harvesting techniques. The following adaptations can be made to maximize the benefits:

- Use of rain gardens and permeable pavements to reduce runoff and promote infiltration.

- Utilizing underground storage systems to prevent evaporation and maintain water quality.

- Incorporating rainwater harvesting systems into new buildings to reduce municipal water usage.

- Utilizing drip irrigation systems for efficient water usage.

It is critical to adapt rainwater harvesting techniques to different climate conditions in order to maximize the benefits of these systems. Individuals and communities can reduce their reliance on municipal water sources, conserve water resources, and promote sustainability by making the appropriate adaptations for different climate conditions.

CHAPTER 10

Rainwater Harvesting for Farming and Agriculture

Nowhere is the necessity of water for life more evident than in farming and agriculture. Many farmers are turning to rainwater harvesting as a sustainable solution in response to rising worries about water scarcity and environmental impact. The process of gathering and storing rainwater for later use is known as rainwater harvesting, and it has many advantages for agriculture and farming.

Rainwater harvesting is a wise and environmentally friendly decision for farmers because it lowers reliance on pricey municipal water sources and increases crop yields. We will delve into the advantages, difficulties, and practical considerations for putting a rainwater harvesting system in place on your farm in this extended chapter as we explore the world of rainwater harvesting for farming and agriculture.

Benefits of Rainwater Harvesting for Agriculture

An age-old practice, rainwater harvesting has recently gained greater relevance than ever. Rainwater harvesting has evolved into a crucial tool for farmers and agriculturalists in response to the challenges of climate change and the rising demand for water in agriculture. In this chapter, we'll look at the many advantages of rainwater harvesting for agriculture, including how it can boost crop yields, lower water costs, and promote a more sustainable future.

1. Reducing Water Costs: The decrease in water costs is one of the main advantages of rainwater harvesting for agriculture. In areas where water is expensive or scarce, harvesting rainwater can significantly reduce the amount of water required for crops. Farmers can reduce their reliance on expensive groundwater and other water sources while saving money on irrigation costs.

Example: A farmer in California had to pay $500 per acre-foot of water from the state water project, but after installing a rainwater harvesting system, they were able to reduce their water bills by up to 60%. This allowed them to save money and invest in other areas of their farm.

2. Conserving Water: Harvesting rainwater is another way to save water and lessen the effects of droughts. Farmers can lessen their reliance on groundwater and other sources of water that may be scarce or depleted by collecting rainwater. The amount of water lost to evaporation and runoff can also be decreased by using rainwater for irrigation.

Example: Farmers in India have been conserving water during droughts for centuries by using conventional rainwater harvesting methods. They were able to irrigate their crops and lessen the impact of droughts on their yields by collecting rainwater in ponds and tanks to use as irrigation.

3. Increasing Yields: Harvesting rainwater can boost yields by giving crops a consistent supply of water. Crops can grow more quickly, more healthily, and yield more when there is a regular and adequate water supply. Additionally free of the minerals and chemicals found in groundwater, rainwater can enhance the quality of crops.

Example: The yields of the farmer's maize and bean crops significantly increased after they installed a rainwater harvesting system on their farm in Oregon. They were able to gather more crops and sell them for more money, which raised their income and enhanced their standard of living.

4. Supporting Sustainable Agriculture: Harvesting rainwater is a sustainable practice that can help with organic farming. Farmers can help preserve natural resources and lessen their negative effects on the environment by reducing their demand for groundwater and other water sources. The use of chemical pesticides and fertilizers can be decreased as a result of rainwater harvesting, which can also improve the health of the soil and encourage biodiversity.

Example: In Texas, a farmer set up a rainwater collection system and began using the collected rainwater for irrigation.

As a result, they were able to use fewer chemical pesticides and fertilizers, which improved the farm's soil quality and biodiversity.

As a result, rainwater harvesting has many advantages for agriculture, including lower water costs, resource conservation, higher yields, and support for sustainable agriculture. Farmers can increase their output, lessen their environmental impact, and help ensure a more sustainable future for agriculture by implementing rainwater harvesting practices.

Different Types of Rainwater Harvesting Systems for Farming

Given the variety of options, picking the best rainwater harvesting system can be difficult. The various types of rainwater harvesting systems for farming, their advantages, and real-world examples will all be covered in this chapter.

- **Above-ground tanks:** These are the most typical systems for collecting rainwater for use in agriculture. They can hold a lot of water and are typically made of steel or plastic. Above-ground tanks are a great option for small to medium-sized farms because they are simple to install and maintain. To provide crops with gravity-fed irrigation, they can be positioned on a raised platform.

- **Below-ground tanks:** Although installed underground, these tanks resemble above-ground tanks. They are a good choice for farms where there is a lack of space or where aesthetics are significant. In addition to offering a cooler water source for irrigation, below-ground tanks can aid in reducing water loss due to evaporation.

- **Ponds:** Ponds are a beautiful and environmentally friendly option for collecting rainwater on farms. They can be utilized for aquaculture, livestock watering, and irrigation. To stop water loss, ponds can be built in a variety of sizes and shapes and lined with plastic. Ponds, however, need regular upkeep to stop algae growth and maintain clear water.

- **Roof catchment systems:** These mechanisms gather water from building roofs and store it in tanks for later use. They are a practical way to collect rainwater in places with a limited amount of space. The amount of water that runs off of buildings can be decreased with the aid of roof catchment systems, which can stop soil erosion.

- **Contour bunds:** Small ridges called contour bunds are built across a slope to help slow down water flow and stop soil erosion. Additionally, they can aid in rainwater collection and enable soil soaking. Contour bunds are an inexpensive and low-tech method of collecting rainwater, making them perfect for small-scale farming.

- **Keyline Farming** - This particular farming technique involves plowing the ground in a way that directs water toward the farm's center. It's been a while since I've done this, but I've been meaning to for a while now. It has been demonstrated that keyline farming increases soil fertility and water retention in the Southwest.

- **Flood Irrigation** - This is a conventional irrigation technique that entails flooding a field with water. After that, the water is allowed to soak into the soil, supplying the crops with moisture. In some regions of the South and the Midwest, flood irrigation is still practiced, but it is becoming less popular as a result of worries about water waste and soil erosion.

- **Rain Gardens** - Rain gardens are made to collect and filter runoff so that it can slowly seep into the ground and replenish the water supply in the soil. They can be modified for use in farming and are frequently used in urban areas. To lessen runoff and enhance soil quality, some farmers in the Northeast have begun incorporating rain gardens into their fields.

How to Maximize the Use of Harvested Rainwater for Crops and Livestock

Rainwater harvesting systems can be highly beneficial for agriculture, providing a reliable source of water for crops and livestock. However, it's important to use harvested rainwater efficiently to get the most out of the system. In this chapter, we will discuss practical ways to maximize the use of harvested rainwater for crops and livestock.

Drip Irrigation

Drip irrigation is among the most effective ways to use rainwater collected for crops. Small amounts of water are directly applied to plant roots using this technique to lessen water loss from evaporation and runoff. Using collected rainwater, drip irrigation systems are simple to set up and can be tailored to different crop types and soil conditions.

Example: A drip irrigation system is used by a Californian farmer to water his almond orchard with collected rainwater. Because of this, he has been able to cut his water consumption by up to 50% while increasing crop yields and lowering water runoff.

Rainwater-fed Livestock Watering Systems

It's possible to hydrate livestock with rainwater collection. In areas with limited access to natural water sources, installing a rainwater-fed livestock watering system may be an affordable way to offer a dependable source of water. These systems can be as straightforward as a rain barrel attached to a trough or as intricate as a gravity-fed system distributing water to numerous locations.

Example: A Texas cattle rancher uses a gravity-fed rainwater harvesting system to supply his animals with water. The system funnels rainwater into a storage tank by collecting it from a number of sizable roofs. The water is then gravity-fed to numerous watering troughs spread out across the ranch from there.

Mulching

Applying a layer of organic material to the soil's surface through the process of mulching aids in moisture retention and reduces evaporation. In addition to enhancing soil health and lowering weed growth, using harvested rainwater in combination with mulching can drastically reduce water consumption.

Example: A vegetable farmer in Oregon waters his crops with a combination of collected rainwater and mulching. By doing this, he has been able to increase soil health and crop yields while reducing his water usage by up to 75%.

Seasonal Planting

Seasonal planting is another way to make the most of the rainwater collected. Crops can benefit from the natural rainfall by being planted during the rainy season, which eliminates the need for irrigation. Optimizing water use can also be accomplished by picking crops that are compatible with the regional climate and rainfall patterns.

Example: When there is a lot of natural rainfall, a farmer in Georgia will plant crops like sweet potatoes and collard greens during the rainy season. As a result, he has been able to cut his water usage by up to 90% while increasing crop yields and cutting costs.

How to Setup Your Rainwater Harvesting for Farming & Agriculture

Step 1: Assess Your Need

Before setting up your rainwater harvesting system for farming and agriculture, you need to assess your needs. Determine the size of your farm and the amount of rainfall in your area. Consider the type of crops you grow, the number of livestock you have, and the amount of water they need. This will help you choose the right system for your needs.

Step 2: Choose the Right System

There are several types of rainwater harvesting systems available for farming and agriculture, including storage tanks, ponds, and dams. Determine which system best fits your needs and budget. Consider the location and terrain of your farm, as well as any legal regulations or zoning laws that may affect your choice of system.

Step 3: Gather the Necessary Equipment and Supplies

Once you have chosen your rainwater harvesting system, gather the necessary equipment and supplies. This may include storage tanks or ponds, pipes, pumps, filters, and any additional tools required for installation. Make sure you have all the equipment and supplies before beginning the installation process.

Step 4: Site Selection

The location of your rainwater harvesting system is crucial for maximum efficiency. Choose a site that is easily accessible and receives plenty of rainfall. Avoid areas that are prone to flooding or have poor drainage.

Step 5: Installation

The installation process will vary depending on the type of system you have chosen. In general, the process involves connecting the pipes and filters to the storage tanks or ponds, and installing the pump if needed. Follow the manufacturer's instructions for installation carefully, or seek the advice of a professional if needed.

Step 6: Maintenance

Regular maintenance is essential to ensure your rainwater harvesting system is functioning properly. Clean out any debris from the pipes and filters, and check for leaks or damage to the system. Consider adding a backup water supply in case of a drought or system failure.

CHAPTER 11

Legal and Regulatory Considerations

Depending on the location, there are many different legal considerations for collecting rainwater. While some states and nations promote the practice, others have strict laws that restrict or outright forbid it. Before installing a rainwater harvesting system, it is crucial to understand and research the local laws and ordinances.

Water rights are one of the main legal factors for rainwater harvesting. In some places, water is regarded as a public resource and cannot be privately owned or managed. Other places permit private ownership and management of water rights, including the right to collect rainwater. To make sure you are in compliance, it is crucial to research the local water rights laws.

Building regulations and licenses are another legal consideration. Rainwater harvesting system installation frequently requires obtaining building permits and adhering to local building codes. The system will be installed safely and in accordance with these regulations. Building codes and permit violations can result in fines or the system being taken away.

Additionally, using rainwater that has been collected may be subject to rules in some areas. For instance, some states forbid the use of rainwater for drinking or cooking, while others mandate specific purification procedures to guarantee the water's safety. To make sure you are using the water safely and legally, it is crucial to research the regulations in your area regarding the use of harvested rainwater.

The effects of rainwater harvesting on the environment should also be taken into account. In some places, excessive rainwater collection can harm the environment by decreasing stream flow or destroying aquatic habitats. Researching the environmental effects of rainwater harvesting in your area is crucial, and you should take precautions to lessen any adverse effects.

Legal considerations for rainwater harvesting

It is essential to find out about any legal implications for your system before you start collecting rainwater. Water rights are an important topic to investigate. You can get in touch with your neighborhood's water authority or department of natural resources to find out whether private ownership and management of water rights are permitted there.

Permits and building codes are additional significant legal considerations. You can ask your local building department or code enforcement agency about the building codes and permit requirements for rainwater harvesting to ensure that your system is installed safely and complies with certain standards.

It's crucial to learn about the rules governing the use of collected rainwater in your area, in addition to building codes and permits. Using rainwater for drinking or cooking may be subject to rules, and the water may also be treated in order to make it safe for consumption. To find out more about these rules, get in touch with your state's or your neighborhood's health department.

The potential environmental effects of rainwater harvesting in your area should also be taken into account. You can speak with your local environmental organization or conservation group to learn whether excessive rainwater collection can harm the environment. You can make sure that your rainwater collection system is both safe and legal by learning about these legal issues, taking the required actions to abide by laws, and minimizing any adverse environmental effects. Here are some real-life practical examples of what a U.S. citizen should do to collect rainwater legally:

- **Research water rights laws in your area:** In some states, such as Colorado, water rights are strictly regulated and individuals are required to obtain a permit to collect rainwater. In other states, such as Oregon, rainwater harvesting is encouraged and there are no restrictions on collection. It is important to research the water rights laws in your area to ensure that you are in compliance. Contact your local water authority or state water resources agency to learn about the specific water rights laws and regulations in your area.

- **Check building codes and obtain permits**: Many states require building permits for the installation of rainwater harvesting systems. For example, in California, a building permit is required for any rainwater harvesting system that is larger than 500 gallons. In addition, the system must comply with state and local building codes. Contact your local building department to find out what permits are required and to ensure that your system meets building code requirements.

- **Learn about regulations on the use of harvested rainwater:** In some states, there are regulations on the use of harvested rainwater. For example, in Texas, harvested rainwater can only be used for non-potable purposes such as irrigation, and it must be clearly labeled as non-potable. In other states, such as Colorado, permits are required to use rainwater for indoor purposes such as flushing toilets or doing laundry. Contact your state or local health department to learn about the specific regulations on the use of harvested rainwater in your area.

- **Consider the environmental impact:** When collecting rainwater, it is important to consider the potential environmental impact. In some areas, excessive collection of rainwater can lead to reduced stream flow and damage aquatic habitats. To minimize any negative effects, consider using a smaller rainwater collection system or incorporating other water conservation practices into your garden, such as using drought-tolerant plants or installing a drip irrigation system.

Regulations in Different States and Regions

Regulations regarding rainwater harvesting can vary widely from state to state and even from region to region within a state. It is important to understand the regulations in your area before installing a rainwater harvesting system to ensure that you are in compliance with local laws. Here are some examples of regulations in different states:

- **California:** In the United States, California has some of the strictest laws governing the collection of rainwater. A permit is typically required to set up a rainwater harvesting system, and a licensed professional must design and install the system. Additionally, only non-potable uses, like irrigation or landscaping, may be made of the rainwater that has been collected. In order to safeguard the state's water supply, California also forbids the collection of rainwater in certain locations.

- **Texas:** Texas has less stringent rules for collecting rainwater than California does. Installing a rainwater harvesting system for residential use does not call for a permit, and the water gathered from the roof can be used for anything, including drinking, cooking, and bathing. However, the homeowner must follow specific water quality standards and the system must be designed and installed in accordance with state laws.

- **Virginia:** Rainwater collection is permitted in Virginia, but localities may have different rules regarding its use. Some regions only permit the use of rainwater for non-potable purposes, while other regions permit its use for potable purposes. To find out the laws in their area, people should contact their local government.

- **Colorado:** Rainwater collection is permitted in Colorado, but there are rules that must be followed. Larger systems that collect more than 110 gallons of water need a permit, and the rainwater they collect can only be used for non-potable indoor uses or outdoor irrigation. A qualified professional must also design and install the system.

- **Florida:** Florida has some rules regarding rainwater collection, but they are generally lax. Residential rainwater harvesting systems that collect less than 2,000 gallons of water are exempt from requiring a permit, and the rainwater can be used for any purpose. However, the homeowner must follow specific water quality standards and the system must be designed and installed in accordance with state laws.

- **Oregon:** Oregon doesn't require a permit for systems that collect less than 5,000 gallons of water to use rainwater. The water that has been collected from the rain can be used for drinking, bathing, and cooking. However, the homeowner must follow specific water quality standards and the system must be designed and installed in accordance with state laws.

- **Georgia:** In Georgia, collecting rainwater for non-potable purposes like landscape irrigation, toilet flushing, and laundry is both permitted and encouraged. However, before installing a rainwater harvesting system, people must get a permit from their local government.

- **New Mexico:** Rainwater harvesting is permitted and encouraged in New Mexico, but before a system is installed, a permit from the local government is required. The state also has laws governing the use of rainwater for potable purposes and mandates that all systems bear labels stating that the water is not fit for human consumption.

Before installing a rainwater harvesting system, it is best to research the most recent regulations in your area because it is important to keep in mind that these rules can change over time. Some local governments might also have additional rules that need to be followed.

How to Ensure Compliance With Local Regulations

To ensure compliance with local regulations when it comes to rainwater harvesting, there are several steps that individuals can take:

- **Research local laws and regulations:** Before installing a rainwater harvesting system, it is important to research the laws and regulations in your local area. This includes both state and local regulations. Contact your state's environmental agency or local water authority to learn about any specific laws or regulations that may apply to your location.

- **Obtain necessary permits:** Many local jurisdictions require permits for the installation of rainwater harvesting systems. It is important to obtain any necessary permits before beginning installation to avoid potential fines or legal issues.

- **Ensure compliance with building codes:** Building codes and regulations vary by jurisdiction and may include requirements for the design, installation, and maintenance of rainwater harvesting systems. It is important to ensure that your system meets all applicable building codes to ensure safety and compliance.

- **Use appropriate treatment methods:** Some states may require specific treatment methods to ensure that harvested rainwater is safe for use. This may include methods such as filtration, disinfection, or other water treatment techniques. It is important to research the regulations in your area and use appropriate treatment methods to ensure that your harvested rainwater is safe for its intended use.

- **Regularly maintain your system:** Regular maintenance of your rainwater harvesting system is essential to ensure its continued function and compliance with regulations. This includes tasks such as cleaning gutters and downspouts, inspecting and repairing any leaks or damage, and regularly testing the water quality.

By following these steps and staying informed about local laws and regulations, you can safely ensure compliance with regulations and legally harvest rainwater for their use in their gardens or homes.

CONCLUSION

As we approach the end of this book, we hope you found the information useful and inspiring. You can not only save water by harvesting rainwater, but you can also create a thriving garden that will add beauty and nourishment to your life.

Remember that every drop of water counts, and by collecting rainwater, you are contributing to a more sustainable future. You will not only be able to grow a lush garden, but you will also be contributing to the planet's health and well-being.

We hope you found this book to be informative and useful on your journey to rainwater harvesting. We encourage you to share your knowledge and experiences with others, as well as to continue to learn and innovate in this field. We can achieve this by putting water conservation and sustainability at the forefront of gardening practices.

Please consider leaving an honest review on Amazon if you enjoyed this book. Your feedback is invaluable and will assist others in determining whether this book is appropriate for them. Thank you for reading and learning about rainwater harvesting for gardeners.

All the best,

Melanie J. Davis

www.ingramcontent.com/pod-product-compliance
Lightning Source LLC
Chambersburg PA
CBHW080607170426
43209CB00007B/1361